PHYSICS
FOR KIDS

49 Easy
Experiments
with Mechanics

Robert W. Wood

Illustrations by Steve Hoeft

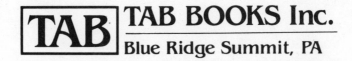

TAB BOOKS Inc.
Blue Ridge Summit, PA

FIRST EDITION
FIRST PRINTING

Copyright © 1989 by **TAB BOOKS Inc.**
Printed in the United States of America

Library of Congress Cataloging-in-Publication Data

Wood, Robert W., 1933-
    Physics for kids : 49 easy experiments with mechanics / by Robert
W. Wood.
      p.   cm.
    Includes index.
    ISBN 0-8306-9282-7   ISBN 0-8306-3282-4 (pbk.)
    1. Mechanics—Experiments—Juvenile literature. I. Title.
QC127.4.W66 1989
531'.078—dc20   89-34646
                                 CIP

TAB BOOKS Inc. offers software for sale. For information and a catalog, please contact TAB Software Department, Blue Ridge Summit, PA 17294-0850.

Questions regarding the content of this book should be addressed to:

Reader Inquiry Branch
TAB BOOKS Inc.
Blue Ridge Summit, PA 17294-0214

Acquisitions Editor: Kimberly Tabor
Book Editor: Lori Flaherty
Production: Katherine Brown

Paperbound cover photograph by Susan Riley, Harrisonburg, VA.

Contents

Symbols Used in This Book

 Protective safety goggles should be worn to protect against shattering glass or other hazards that could damage your eyes.

 Materials or tools used in this experiment could be dangerous in young hands. Adult supervision is recommended.

 Electricity is used in this experiment. Young children should be supervised and older children cautioned about the hazards of electricity.

 Scissors are used in this project. Young children should be supervised carefully and older children instructed to exercise caution.

 Exercise caution around any open flame. Adult supervision recommended.

 Use **extreme** care when handling a razor blade. Adults should supervise young children **very** carefully. Place razor blade out of reach after use.

INTRODUCTION

Physics is the science that explores the natural world around us. It tells us how and why a lever can lift a heavy weight, why hot air rises, and what light is. It is the study of electricity and magnetism, and how sound waves travel. This fascinating science covers such a wide range of subjects that no simple definition explains it.

To better understand physics, it is broken up into smaller fields: mechanics, heat, light, electricity and magnetism, and sound. But sometimes these fields overlap each other. For example, a student studying electricity and magnetism to learn why a telephone worked would also learn how sound waves vibrate objects that send electrical signals.

It was probably prehistoric man that discovered he could skip a flat rock across a body of water, and everyone knows that by lowering the air pressure in a straw, you can make a soft drink or milkshake flow from a glass.

This book is an introduction into the exciting world of physics. The subject of this book is mechanics. It was Isaac Newton that first described building and using machines as mechanics. Engineers use mechanics to make sure bridges stand up under the stress of the loads they will carry. Mechanics is used to design airplanes, rockets, and space vehicles. Scientists use mechanics to study the motion of atomic particles. Astronomers use the principles of mechanics to determine the movement of planets and stars, and certainly, it's important to the success of orbiting satellites and the space shuttle. Physics is the study of the effects of forces on bodies or fluids at rest, or in motion. This book is divided into two parts, fluid mechanics and solid mechanics.

Science experiments can be fun, but be careful. Safety should always be the first consideration.

Carefully look over the symbol key, *Symbols Used in This Book*, at the beginning of the book before beginning any experiment. These symbols mean that you should use extra safety precautions, or that some experiments might require a parent or teacher's help. *Always* refer to the key whenever you see a warning symbol before proceeding.

Part I

FLUID MECHANICS

Fluid mechanics is the study of how forces and motions affect fluids and gases. It includes the study of fluids at rest, called hydrostatics; fluids in motion, called hydraulics; and the study of air moving around objects, called aerodynamics. The following experiments provide the basics for understanding fluid mechanics.

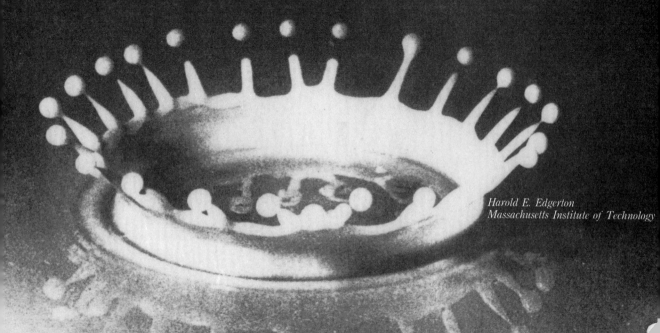

Harold E. Edgerton
Massachusetts Institute of Technology

Experiment 1

Needle Floating
on Water

Materials
- [ ] bowl of water
- [ ] needle
- [ ] fork
- [ ] liquid detergent

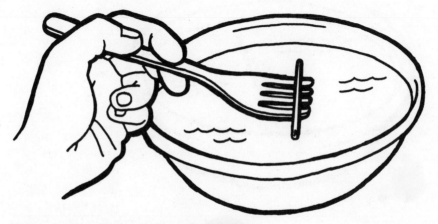

Surface tension allows a needle to float on water.

Using the fork, carefully lower the needle to the surface of the water. Slowly remove the fork and the needle will float. It is supported by surface tension. Surface tension is when "like" molecules form a cohesive bond. Cohesive bonds are formed when the molecules of a part are attracted, and the molecules come together. The molecules press together, forming a thin skin. This thin skin, called surface tension, is surprisingly strong and will support objects that would normally sink. Now pour a drop of detergent into the water. The needle will sink. The detergent lowers the surface tension and the skin becomes weaker and will not support the needle.

A drop of liquid soap will make the needle sink.

Experiment 2

Water Supported by Wire Basket

Materials

- ☐ small tea strainer
- ☐ cooking oil
- ☐ water

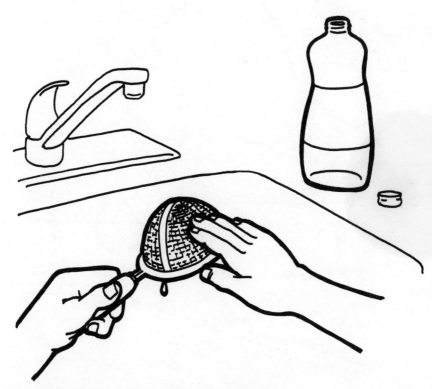

Cooking oil helps cover any sharp points in the screen.

Over a sink, coat the wires of the strainer with oil. Shake the strainer to remove any excess oil. The holes in the wire should be open. Putting oil on the wire forms an adhesive bond between the wire and the oil. It also makes the size of the wire mesh smaller.

Adhesive bonds are formed when "unlike" molecules are attracted and press together.

Slowly pour water down the inside edge of the strainer. The strainer will fill with water. Because the mesh size is smaller, the adherence of the water to the oil and wire is strong enough to support the weight of the water.

If you touch the wire with your finger, the water will run out. Your finger has a greater force of attraction to the water in the strainer than the oil. Also, some oil is attracted to your finger. Both of these forces are due to adhesion.

Shaking the screen removes excess oil and opens the holes.

The strainer will now hold water.

Touching the screen breaks the adhesive forces.

Experiment 3

Water through Handkerchief

Materials
- [ ] water
- [ ] wide mouth jar or drinking glass
- [ ] rubber band
- [ ] a handkerchief

The cloth should be stretched tightly over the mouth of the jar.

Over a sink, place the handkerchief over the top of the jar or glass and fasten it in place with a rubber band. Pour water through the handkerchief into the jar, filling it to the top. Over a sink, carefully turn the jar upside down. The water does not run out.

Use a slow stream of water to fill the jar.

The water can be poured in because the surface tension is broken by the force of the water being poured. The water does not run out because the water in the tiny holes in the cloth formed a surface adhesion, and the air pressing up against the cloth is equal to the pressure inside the jar.

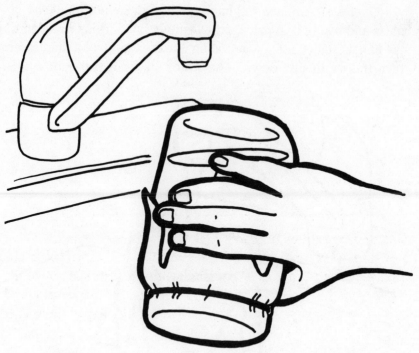

Air pressing up against the cloth, and surface adhesion keeps the water from running out.

Experiment 4

The Oil Spot

Materials

- ☐ soap
- ☐ water
- ☐ drinking glass
- ☐ toothpick
- ☐ cooking oil

Pour a few drops of oil on the top of a glass of water. Oil and water do not mix and the surface tension of the water will make the oil move to form a round spot. Use the toothpick to move the oil spot over the surface of the water. The spot can be bounced off the sides of the glass as it tries to return to the center of the water.

Oil will form a round spot on calm water.

Place a small amount of soap on the tip of the toothpick and gently touch it to the center of the oil spot. Instantly the oil will spread in a wide circle and move to the sides of the glass. The soap formed a very thin film that expanded in a wide circle, pushing the oil outward to quickly form the large ring.

Oil molecules cling together and can be easily moved with a toothpick.

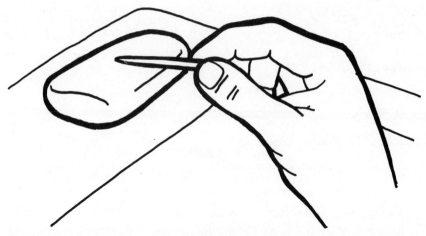

It only takes a small piece of soap on the tip of a toothpick to break surface tension.

The oil quickly rushes to the edge of the water.

Experiment 5

Stretched Water

Turn on the cold water faucet in the kitchen sink enough to get a smooth flow of water. Hold the end of the spoon in the water, about half way down the flow. When the center of the cup-part of the spoon is in the center of the flow, a round sheet of water will form. Air pressure that causes the surface tension holds the water together to form the sheet. Even when the water breaks up into droplets, the surface tension is still working, making it fall in round drops.

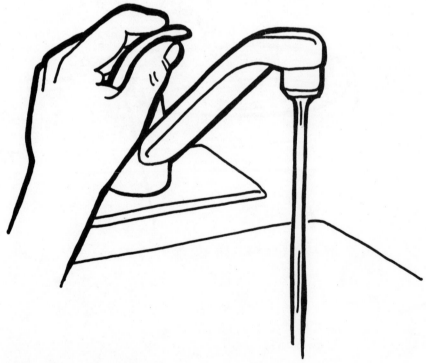

Surface tension tries to hold the water in a smooth flow.

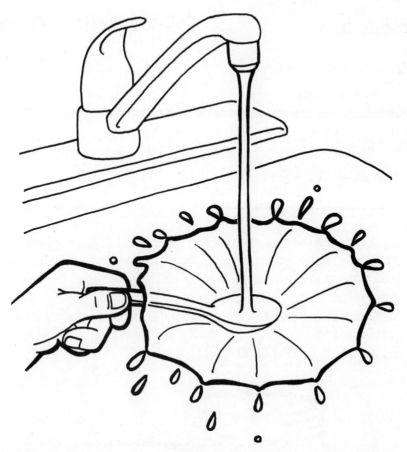

A spoon will spread the flow, but surface tension still holds the water together.

Experiment 6

Water Filter

Materials

- ☐ 2 jars, one taller than the other
- ☐ colored water or tea
- ☐ length of paper towel

Fill one of the jars about three-fourths full of the colored water. Twist the paper towel into a roll and put one end in the jar of colored water and the other end in the empty jar. The experiment will work faster if the roll of paper towel is first completely wet, and the empty jar is lower than the jar of water. With the empty jar lower, gravity will speed things up. Water will then begin to drip into the empty jar. It should be almost clear.

Muddy or colored water can be filtered from one jar to another.

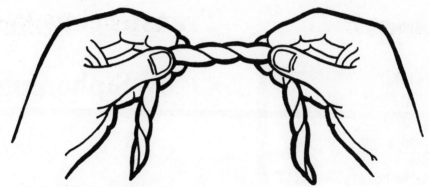

A paper towel or cloth towel can be used as a filter.

The water travels from one jar to the other because of capillary attraction. This is caused by adhesion. The water molecules are attracted to the molecules in the paper towel. As the water moves through the paper towel, it is gradually filtered.

Air pressure and gravity make the filter work.

Experiment 7

Transferring Water by Siphoning

Materials

- [ ] two jars, about the same size
- [ ] length of rubber tubing
- [ ] water
- [ ] chair or box

Fill one of the jars about three-fourths full of water and place it on a table. Put the empty jar on a box or chair next to the table. Fill the tube with water and pinch both ends to keep the water in. Lower one end of the tube in the jar with the water, and stick the other end in the empty jar on the box or chair. Release the pinched ends and water will start to flow into the empty jar. This continues until the water has moved from the higher jar to the lower jar, or the end of the tube is removed from the higher jar. This happens because gravity causes the water to flow through the tube. This lowers the pressure inside the tube. The air pressure pushes on the top of the water in the upper jar, forcing more water in the tube to equal the lost pressure. A siphon is a tube that uses gravity and air pressure to transfer water.

The jar of water must be higher than the empty jar.

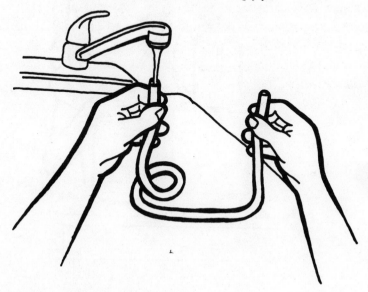

The tube should be completely full.

A little practice may be necessary to get the water to flow.

The water will flow until air enters the higher end of the tube.

Experiment 8

Weight of Different Fluids

Pour a small amount of the colored water into the glass or jar. About one inch deep will do. Tilt the glass and slowly pour in a layer of cooking oil. Pour down the inside edge of the glass so the fluids won't mix. Next, pour in a layer of rubbing alcohol. You should have

Water, oil, and alcohol are a few of the fluids that don't weigh the same.

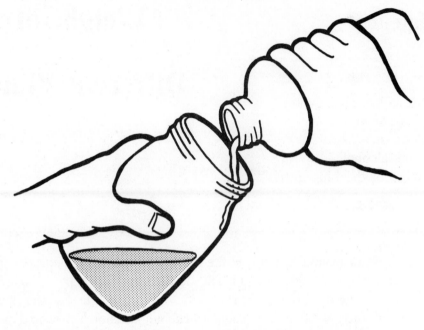

Pouring the fluids down the side of the jar helps keep them from mixing.

three separate layers of fluids. The heaviest fluid at the bottom, the next heaviest above, and the lightest at the top. Fluids will arrange themselves according to their weight. In this case, water is the heaviest.

Fluids of different weights find their own level.

The weight of the coin, and a wet bottle top, will form a seal.

Experiment 9

Moving D

Materials

- ☐ dime
- ☐ refrigerator
- ☐ empty pop bottle

Place the empty bottle in the refrigerator for 10 or 15
Remove the bottle and wet the top a little to form a seal
dime. Place the dime on top of the bottle and wrap your hand
the lower half. After a few minutes, the dime will start to
fall back. Start to lift again, and again, fall back.

Use a refrigerator to lower the air pressure in the bo

The warmth from your hands increases the pressure inside the bottle.

This happens because cold air is dense and takes up less space than warm air. At first, the dime on the moistened rim of the cold bottle made a seal. But as your hands warmed the bottle and the air inside, the warm air expanded and created a little pressure inside the bottle. This pressure lifted the dime, allowing some of the air to escape. The dime then falls back into place. This continues as long as your hands are able to warm the air inside the bottle.

Experiment 10

Air Takes Up

Space

Place the balloon about half way into the jar and blow up the balloon. The balloon will inflate but does not completely fill the inside of the jar. Release the air from the balloon and try it again. This time place the straw in the jar. The balloon will inflate and completely fill the jar.

In the first attempt, the balloon expanded and the air between the balloon and the glass was trapped. This trapped air exerted a pressure on the balloon equal to the pressure of the balloon on the air. As a result, the balloon expanded outside of the jar, where there was less opposition. When the straw was stuck in the jar, it allowed the air to escape and the balloon to fill the jar.

A straw keeps the mouth of the jar from being sealed.

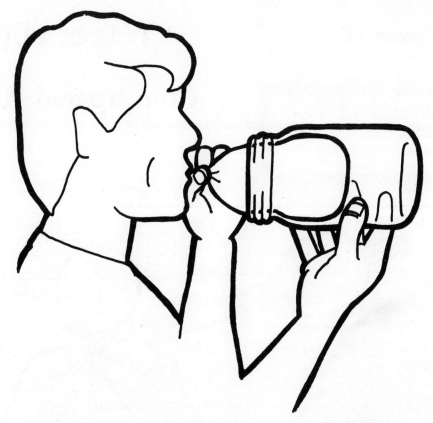

Air trapped inside the jar keeps the balloon from fully expanding inside.

Now there is no trapped air in the jar.

Experiment 11

Weight of Compressed Air

Materials

☐ yardstick
☐ two balloons
☐ needle

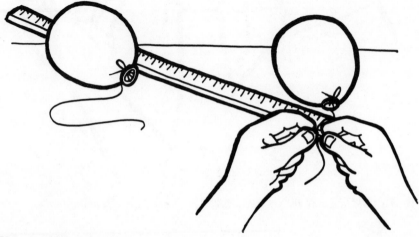

Use a yardstick for balance.

Inflate two balloons to about the same size and tie a length of string to each one. Suspend the yardstick from the center and balance the balloons on each end. Puncture one of the balloons with the needle. The inflated balloon will upset the balance. This proves that compressed air weighs more than normal air.

The weight of the compressed air in each balloon is balanced.

The popped balloon has upset the balance.

Experiment 12

Measuring Warm Air and Cold Air

Materials
- ☐ tape measure
- ☐ balloon
- ☐ refrigerator

Inflate a balloon and tie the end so that no air can escape. Measure the balloon around the middle with the tape measure and write down the reading. Place the balloon in the refrigerator for about half an hour. Remove the balloon and measure it again. Compare the two measurements. The balloon is smaller. When air molecules are cooled, the air takes up less space.

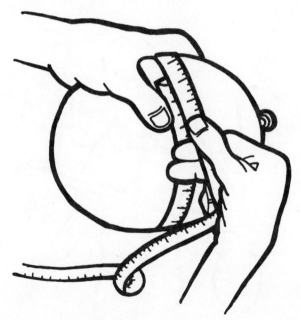

Use a cloth tape measure to measure the balloon.

The refrigerator cools the air in the balloon, causing it to get smaller.

Experiment 13

Lifting a Heavy Object with Air

A balloon can be used to lift books.

The balloon contains the air pressure and lifts the books.

Place the balloon on the edge of a table and put the books on top of the balloon. Slowly inflate the balloon and watch the books rise. Air pressure from the lungs is strong enough to lift the books. This is the principle of hydraulics. This is how a tire pump can be used to inflate a tire and lift a heavy automobile.

Experiment 14

Hot air Balloon

The ice water cools the air in the bottle.

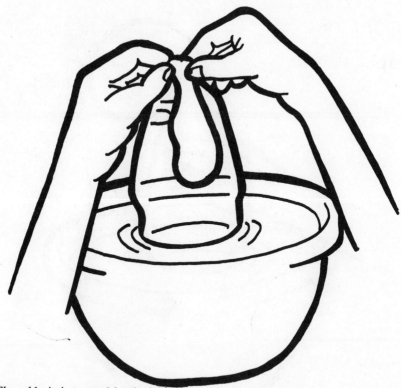

The cold air is trapped by the balloon.

Place an empty bottle in a bowl of ice water for about five minutes. Fit the end of a balloon over the top of the bottle and fasten it with the rubber band. Remove the bottle from the ice water and allow it to warm a little. Next, place the bottle in the bowl of hot water. The balloon will begin to inflate. This is because the air in the bottle was cooled by the ice water. The colder air took up less space. When the cold air was trapped inside the bottle by the balloon and the air inside the bottle was heated by the hot water, it began to expand and inflated the balloon. Warmer air takes up more space than normal air. This makes it less dense, so it will rise. This is how hot air balloons are able to fly.

The warm air expands and inflates the balloon.

Experiment 15

Balloon Jet

Materials

- [ ] balloon
- [ ] drinking straw
- [ ] long string (10 to 15 feet)
- [ ] scotch tape
- [ ] two chairs

Place a chair on each side of the room and feed one end of the string through the straw. Tie each end of the string to each chair. Blow up the balloon and tape it to the straw in a couple of places. Put the balloon near one of the chairs. When the air in the balloon is released, the balloon will speed along the string to the other chair.

This happens because every action has an equal and opposite reaction. The air pressure in the balloon produced an exhaust from one end. This exhaust pushed against the stationary air in the room creating a force called thrust. The thrust caused the balloon to move forward. This is the principle that allows jet airplanes to fly.

The string must slide freely through the straw.

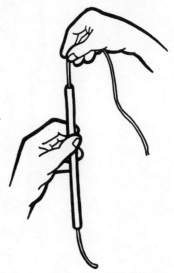

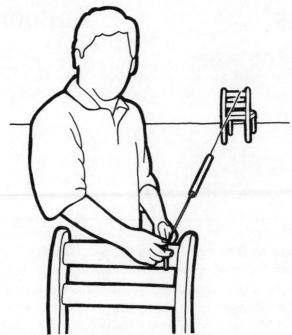

The string should be stretched tight between the chairs.

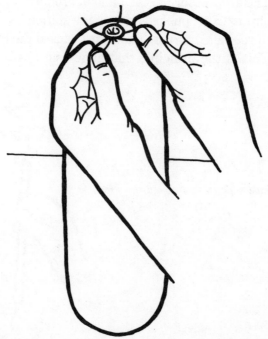

The mouth of the balloon can be pinched closed or tied so it can be easily untied.

38

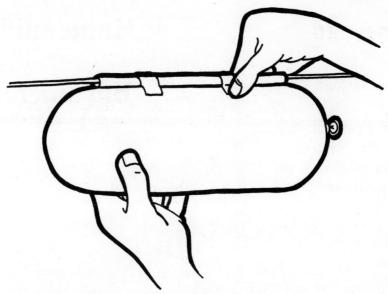

Use scotch tape to hold the balloon to the straw.

Experiment 16

Homemade Barometer

Materials

- ☐ book
- ☐ rubber band
- ☐ piece of thin rubber (busted balloon)
- ☐ wide mouth jar
- ☐ toothpick
- ☐ glue
- ☐ drinking straw
- ☐ cardboard

The lightweight pointer can be made from a toothpick glued to the straw.

A barometer is an instrument used to measure atmospheric pressure. This is useful in monitoring changing weather patterns to forecast the weather. A type of barometer, calibrated in feet, is used in aircraft to tell the height of the airplane above sea level.

Use a small amount of glue to attach the other end of the straw.

Stretch the piece of rubber across the top of the jar and fasten it in place with the rubber band. Glue the toothpick to the end of the straw to make a pointer then glue the other end of the straw to the center of the rubber cover on the jar. Make a scale on the cardboard and place the end in a book. You can mark your scale from your local weather report. Use the rubber band to hold the book closed. Next, position the jar and pointer near the book and scale so that the end of the toothpick points to the scale. Place the barometer where the temperature stays about the same, away from drafts, heaters, and direct sunlight.

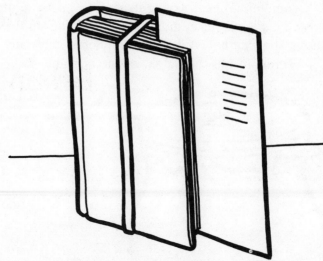

A scale is necessary to show the changes in air pressure.

The pointer will move up and down the scale when the air pressure outside the jar varies with the pressure inside the jar. If a high pressure weather system moves in, the increased pressure will press in the rubber cover and move the pointer up the scale. A low pressure system will move the pointer down the scale.

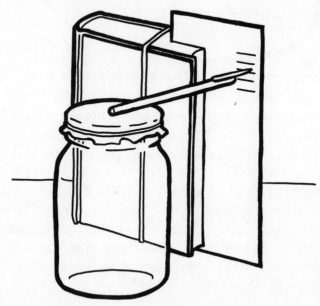

The pointer will move when the air pressure in the room increases or decreases.

42

Experiment 17

Diving Medicine Dropper

Materials
- ☐ medicine dropper
- ☐ empty plastic bottle
- ☐ water

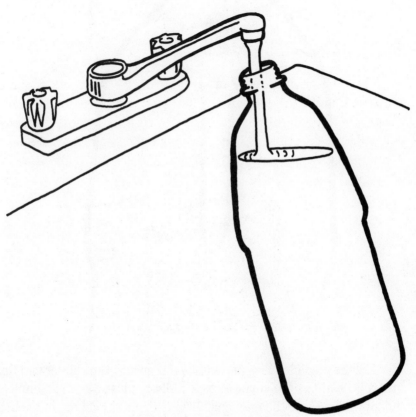

A plastic bottle filled with water is the diving chamber.

The top of the medicine dropper should float just above the surface.

Fill the medicine dropper with water so that the rubber end floats just a little above the surface. Next, fill the plastic bottle completely full of water. Place the medicine dropper, rubber end up, in the bottle and screw the cap in place. The medicine dropper will float at the top until the bottle is squeezed. Then the medicine dropper will dive to the bottom. With gentle squeezing, the medicine dropper can be made to float at any level in the bottle.

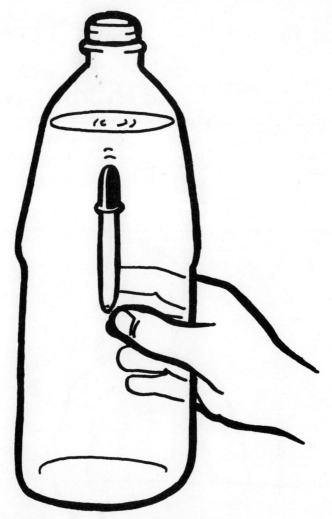

Squeezing the bottle causes the medicine dropper to dive.

This happens because water is very hard to compress, while air can be compressed easily. When the plastic bottle is squeezed, the water is not compressed, but the air inside the medicine dropper is. This allows more water to enter the medicine dropper and it sinks. When the pressure on the bottle is released, the compressed air in the medicine dropper pushes some of the water out and the medicine dropper rises.

Experiment 18

Air Pressure

Holding a Stick

The yardstick is over the edge of the table.

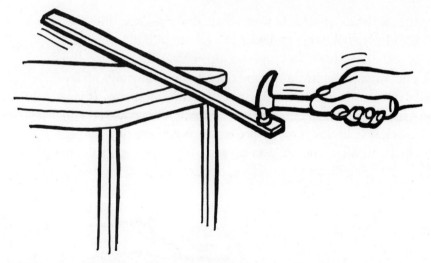

Air presses on the length of the yardstick equally.

Use your hands to smooth the newspaper.

Place the yardstick on a table with about 12 inches sticking over the edge. Carefully strike this end with the hammer. The other end will pop up off the table. Now spread the newspapers over the yardstick and smooth them flat against the table. Carefully strike the exposed end again. This time the other end does not rise. If you strike the exposed end hard enough, it will break. The covered end of the yardstick is held down by air pressure. When the newspaper was smoothed out, all of the air was pressed from beneath the paper. The air, pressing down from above, kept the stick from rising.

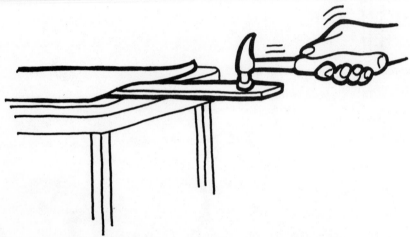

Now there is much more air pressure pressing on top of the yardstick.

Experiment 19

Boiled Egg in a Bottle

Carefully drop a burning match or small wad of burning paper into the milk bottle and, just before the flame goes out, place the egg, with the smaller end down, in the opening of the bottle. The egg is pulled inside the bottle. To get the egg back out, turn the bottle upside down, with the smaller end of the egg in the opening, and blow hard into the bottle. Stop blowing suddenly and the egg will move out of the bottle. It is not always possible to generate the pressure needed to force the egg back out.

The small flame inside the bottle heated the air and caused it to expand. Some of it escaped out of the mouth of the bottle. The egg sealed the opening in the bottle, then the flame went out, and the air inside the bottle began to cool. As the air inside cooled, it contracted and took up less space. The outside air has more pressure and pushed the egg into the bottle. Blowing into the bottle increased the inside air pressure and forced the egg back out.

The flame causes the air to expand.

The boiled egg seals the mouth of the bottle.

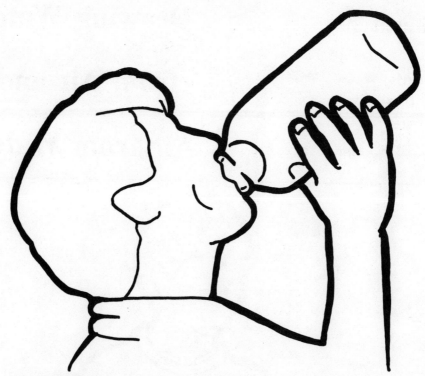

Blowing the bottle increases the air pressure inside.

Experiment 20

Drawing Water from Air and Air from Water

Materials

- [ ] two drinking glasses
- [ ] ice water

Use ice water to fill the glasses.

A window sill is a good source for sunlight.

Fill two glasses with ice cold water. Place one near a heat source, or in direct sunlight. Put the other one in normal room temperature. Let the two glasses stand for several minutes. Soon, both glasses will feel damp on the outside and beads of moisture might appear. The glass in direct sunlight, however, should have air bubbles clinging to the side of the glass.

The cold glass cools the air next to it. This causes the warm air to release its moisture and deposit it on the outside of the glass. The other glass has bubbles sticking to the side of the glass because cold air has tiny molecules of air that expand and form bubbles when the cold water is warmed.

Warm air holds more moisture than cold air. This explains why we have warm muggy days in the summer but never have cold muggy days in the winter.

Sunlight will cause air bubbles to appear.

Experiment 21

Measuring Dew Point

Materials

- ☐ thermometer
- ☐ tin can
- ☐ water
- ☐ ice
- ☐ salt

Add ice to the water in the can.

The dew point is when moisture begins to appear on the can.

Dew point is the temperature that moisture in the air condenses and forms water droplets.

Pour water in the tin can and add ice, slowly stirring. Place the thermometer in the can and watch for when moisture begins to form on the outside of the can. At that instant, the temperature is the dew point. This temperature can vary and depends on the amount of moisture in the air.

If you continue to stir and add salt and a little more ice, the moisture on the can will freeze and change to frost. Salt lowers the melting temperature of ice.

Experiment 22

Water Pouring from a Jug

Water will easily flow into a jug by pushing air out the top; but will not flow out as easy.

Over a sink, fill the jug about half full of water. Next, turn it upside down. The water will gurgle and the jug will gulp air as the water flows from the jug. Notice how long it takes to pour out the water. Try it again, but this time, stick the straw in the jug so that one end is above the water. This time the water will flow more smoothly and quickly from the jug.

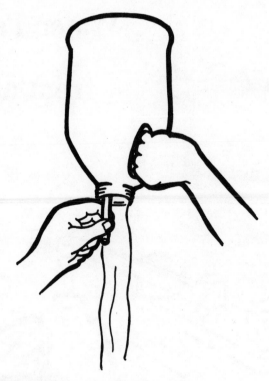

A straw lets air in to fill the space above the water.

Gravity pulls the water from the jug. But when the water starts to flow, the air pressure inside the jug decreases and the air pressure in the room tries to keep the water in. Then, as air bubbles rise through the water equalizing the air pressure, water will flow in gulps from the jug.

When the straw is inserted, air flows through the straw and keeps the inside and outside air pressure the same and the water flows smoothly.

Experiment 23

The Bernoulli Effect

Materials
☐ glass of water
☐ drinking straw
☐ single edged razor blade

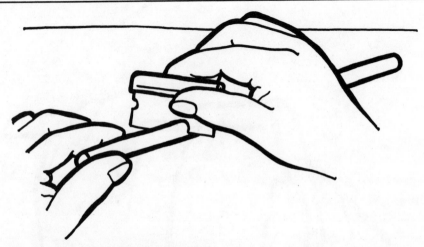

Cut the straw across with the razor blade.

Daniel Bernoulli (1700-1782) was a Swiss mathematician who discovered that the pressure of air is lowered as the speed of the air is increased. This principle can be demonstrated by building a simple atomizer or sprayer.

Carefully cut a slit crossways in the straw about two inches from one end. Be sure not to cut all the way through. Flatten the longer part of the straw and bend it at the cut. Put the short end in the glass of water so that the bend is just above the surface of the water and at the far edge of the glass. Blow hard through the straw and a mist of water will spray from the opening in the bend.

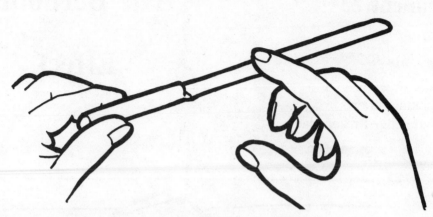

Use your finger to partially flatten the long part of the straw.

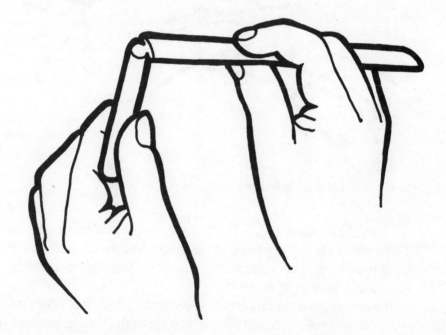

Bend the short end of the straw down.

This happens because the jet of air through the opening lowers the pressure. Then the normal air pressure pressing down on the surface of the water forces it up through the short end of the straw and it is blown away.

The water moves up the tube and is sprayed away.

Experiment 24   Specific Gravity

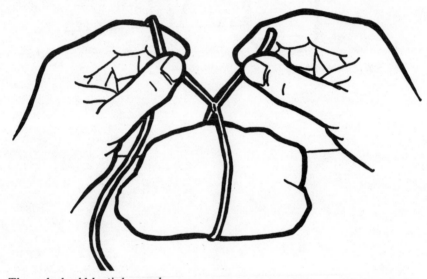

The rock should be tied securely.

The specific gravity of an object is determined by how its density compares with the density of water. For example, gold has a specific gravity of 19. This means that any volume of gold is 19 times heavier than the same volume of water.

According to legend, the Greek mathematician Archimedes (287BC-212BC) discovered this principle while trying to find out if King Hiero's royal crown was made of pure gold.

First, weigh the rock in air then lower it into the water and compare the weight. The specific gravity is the weight in air divided

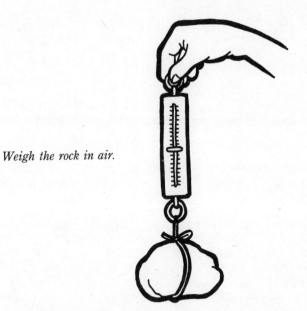

Weigh the rock in air.

by the amount of weight it lost in the water. For example, if the rock weighs 5 pounds in air and 3 pounds in water, subtract 3 from 5, which means the rock displaced 2 pounds of water. Divide the weight in air by the weight of the water it displaced, 5 divided by 2, to find the specific gravity of the rock to be 2.5.

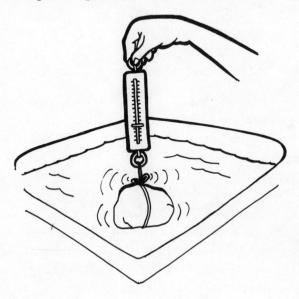

Weigh the rock under water.

Part II

SOLID MECHANICS

Solid mechanics is the study of dynamics, or the study of statics. Dynamic means energetic, vigorous, or forceful. In physics, it deals with the forces affecting moving objects or bodies. The energy of a moving object is called kinetic energy. Statics deals with bodies, masses, or forces at rest. The experiments that follow demonstrate some of the basic principles of solid mechanics.

Simple Machines

Normally, we think of a machine as something with a motor that performs work, which it is. But a machine is really any device that does work. It doesn't have to have a motor. A machine produces a force and controls the direction and speed of the force, but cannot produce energy. It can never do more work than the amount of energy put in to it because of the friction of parts. The lever is probably the most efficient. The work it puts out nearly equals the energy put in because the energy lost to friction is very small.

Experiment 25

Inclined Plane and Wedge

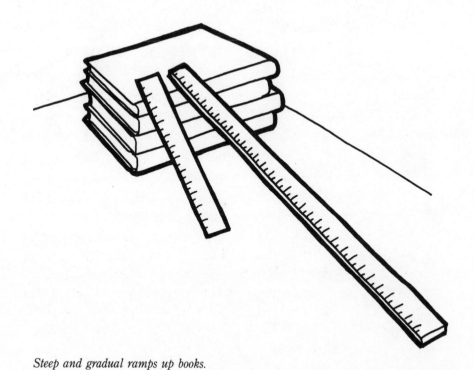

Steep and gradual ramps up books.

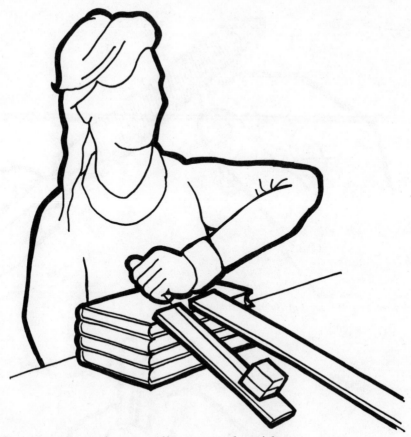

The steep ramp requires more effort to move the weight.

Stack the books about 6 inches high and make two ramps up the books with the rulers. Attach the rubber band to the weight and pull it up the ramps. Notice how far the rubber band stretches. The steeper the ramp, the more the rubber band stretches.

The ramp is an inclined plane. It is a machine that makes it easier to gradually move a load to a height instead of lifting it straight up. The longer the slope, the less effort required, however, the distance traveled is longer, so the amount of work is the same. This is why roads up a mountain are winding instead of going straight to the top.

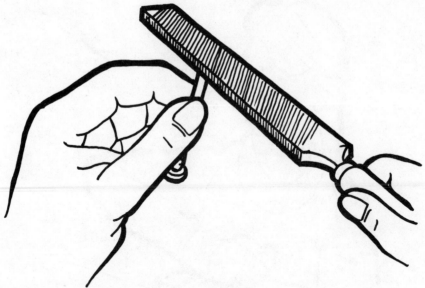

Removing the point of the nail removes its wedge.

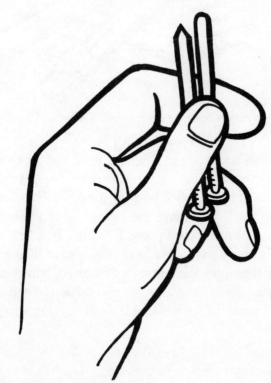

One nail with a wedge, one without.

The nail without the wedge is harder to drive.

File the point of one of the nails so it is blunt and hammer it into the block of wood. Now hammer in the nail with the point. Notice how much easier the pointed nail goes in. This is because the end of the nail is a wedge. A wedge is made by putting two ramps or inclined planes together. The sloping point on the nail makes the penetration more gradual so it takes less effort. Axes, chisels, and needles are some examples of wedges.

Experiment 26

Screw and Lever

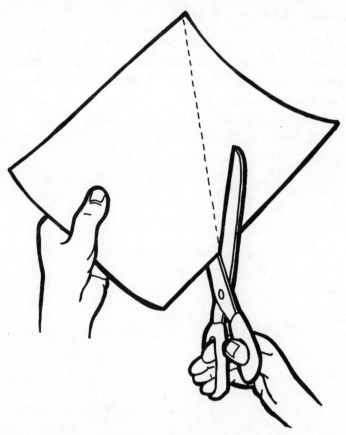

A ramp cut from paper.

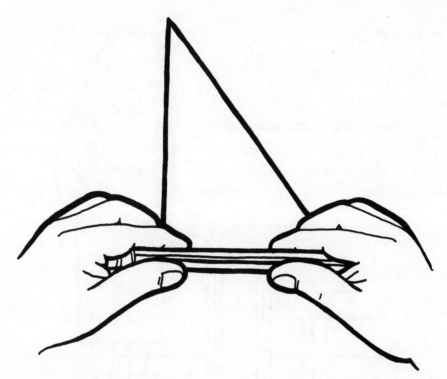

A ramp twisted into a tight circle.

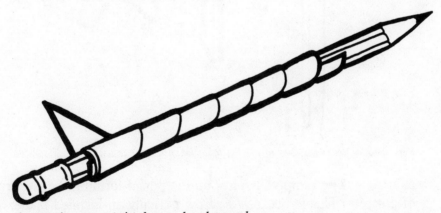

A screw is a ramp twisted around and around.

 Cut a right triangle from the sheet of paper to make a ramp. Roll the paper on the pencil from the short side of the triangle to the point. Use a marker to color the cut edge. This will make the spiral stand out. Keep the bottom, or base line, of the triangle even

Very little effort is required to lift the table when a lever is used.

as it rolls. The ramp will spiral up the pencil forming the pattern of a screw. This is because a screw is really an inclined plane.

A lever is a stiff bar that turns around a pivotal point called a fulcrum. The advantage is in the short distance between the load and the fulcrum, and a long distance between the fulcrum and the point where the effort is applied.

To make a lever, place one of the boards on its end next to the table, and the other board on top. With one end of the board under the edge of the table, press down on the other end, the heavy table can be easily lifted.

Some common levers include a bottle opener, see-saw, wheelbarrow, hammer, pry bar, and can opener.

Experiment 27

Wheel and Axle and the Pulley

Materials

- ☐ pencil sharpener
- ☐ string
- ☐ 2 or 3 books
- ☐ wire clothes hanger
- ☐ empty thread spool
- ☐ weight (washer or fishing sinker)
- ☐ wire cutter

Using a crank to turn a wheel and axle makes lifting a load easier.

The wheel and axle really operate on the principle of the lever. The center of the axle is the pivotal point, or fulcrum. The mechanical advantage depends on how large around the axle is and the distance traveled by the crank handle where the effort is applied.

The pulley is a form of the wheel and axle. With only one pulley, the force to lift an object is the same as the weight of the object. The only advantage with one pulley is that you can pull from a different direction and add your weight to the force of the pull.

Tie one end of the string around the center of the books. Remove the cover from the pencil sharpener and tie the other end of the string around the end of the shaft. When the handle is cranked, the string will wind around the shaft and the books can be easily lifted. The pencil sharpener is used as the wheel and axle.

To make a simple pulley, cut the bottom part of the clothes hanger. Bend the ends and thread them through the holes in the spool.

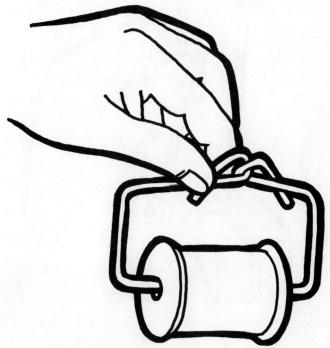

A simple pulley can be made from a spool and piece of wire.

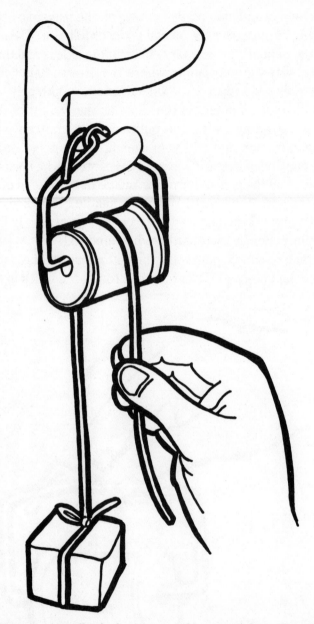

The advantage of a pulley is that you can add your weight to the pulling force.

Bend the ends together to keep the wires from separating. Attach the pulley to a support and, using a string and a small weight, you can see that the downward force used to lift a weight is the same as the weight itself.

Experiment 28

Static Friction

and

Moving Friction

Materials

- [ ] several rubber bands
- [ ] shoe box with something inside for weight
- [ ] three pencils

When an automobile begins moving, it uses a transmission to go from the lower gears to the high gear it uses to travel at highway speeds. This means that it takes more force to start something moving than it does to keep it moving.

Connect the rubber bands together and attach one end to the shoe box. Place the box on a smooth floor or table and pull on the other end of the rubber bands. See how far they stretch before the box starts to move. Then notice how far they are stretched to keep the box moving. The rubber bands are stretched farther to get the box moving. This is because static friction is greater than moving friction.

Now place the three pencils under the box and try the static test again. It will take much less force to start the box moving. Roller bearings are used to reduce friction.

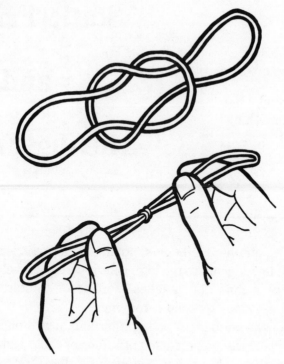

Rubber bands are looped together instead of tying.

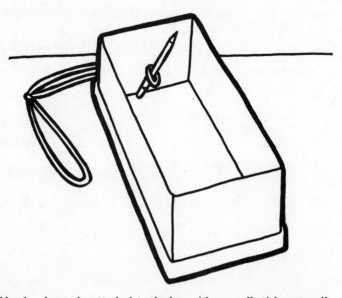

The rubber bands can be attached to the box with a small stick or pencil.

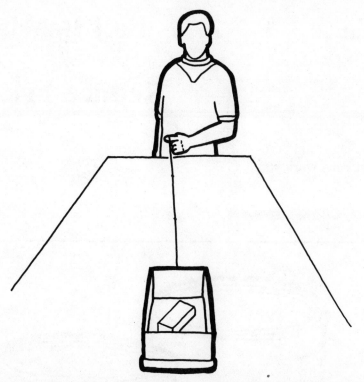

How far the rubber bands stretch demonstrate how much effort is required.

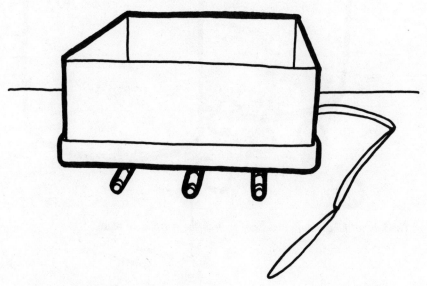

Pencils act as bearings to reduce friction.

Experiment 29

Ball Bearings
Reduce Friction

Materials
☐ marbles
☐ 2 tin cans, the same size, with grooves around the top (paint cans work well)
☐ book

Grooves in the cans are necessary to keep the marbles in place.

The marbles reduce the amount of contact between the two cans.

Place the marbles in the groove in one of the cans. Invert the other can and place it on top so that the groove is over the marbles. Place the book on top and notice how easily the can can be turned. Rolling friction is much less than sliding friction. Ball bearings are hard and very little contact is made with the surface. Also, ball bearings are really rollers, which are not limited in their direction of rotation.

The marbles act as bearings, allowing the top can to turn.

Experiment 30

Balancing Act

Cut a notch in one end of the match and push the other end in the center of the bottom of the cork. Push the forks in opposite sides of the cork with the handles pointing the same direction as the match.

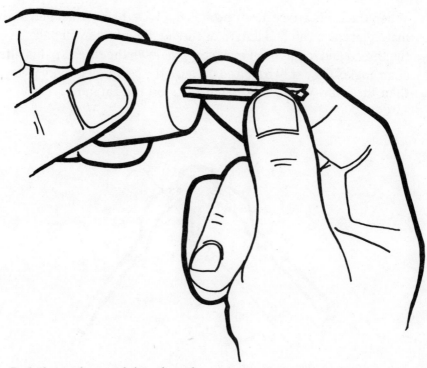

Push the wooden match into the cork.

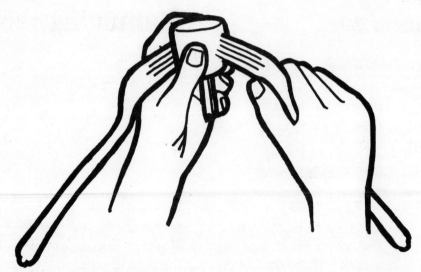

Insert the two forks into the cork.

When the forks are securely attached, place the notched end of the match on the stretched length of thread. The forks will balance on the thread and might even move up and down the thread if it is tilted at an angle. They stay balanced because the center of gravity is lower than the point where the match rests on the thread.

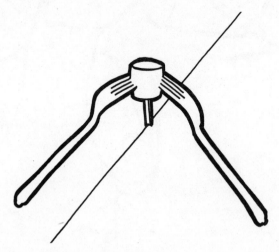

The forks can be balanced on a thread.

Experiment 31

Defying Gravity

Materials
- [ ] ruler
- [ ] hammer
- [ ] short length of string
  (about 10 or 12 inches)

Tie the string to form a loop about 4 or 5 inches across. Slip the loop about two-thirds of the way down the ruler, and about half way down the handle of the hammer. Place the tip of the ruler on the edge of a table. This should put the head of the hammer slightly under the table, and the end of the handle about one-third of the way down the ruler. It might take a little adjusting. The ruler and hammer balance, with the very tip of the ruler on the edge of the table, because the center of gravity has been positioned directly below the edge of the table.

The center of gravity is that one point where all of the mass of an object seems to exist. Because gravity acts on mass to create weight, this one point is the center of gravity, or the point where gravity acts on the object to create the object's weight. If an upward force is exerted at this point, equal to the object's weight, it will be balanced.

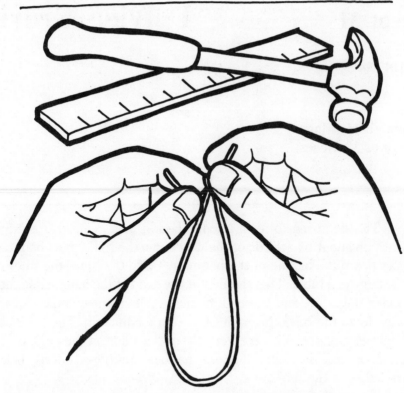

A ruler, hammer, and loop of string can perform a balancing act.

The loop of string is fitted around the ruler and the hammer handle.

88

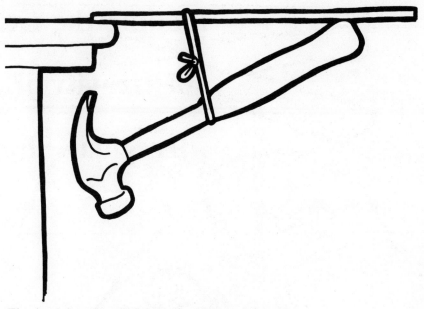

The tip of the ruler will balance on the edge of the table.

Experiment 32

Finding the Center of Gravity of an Irregular Shape

Materials

- [ ] piece of cardboard in an irregular shape
- [ ] string with a weight (washer)
- [ ] pencil

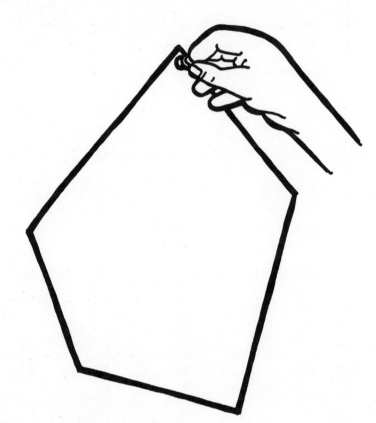

Suspend the cardboard by one of its corners.

Hold the piece of cardboard from one of its corners. Attach the string with the weight to this point. Mark a line along the string near the center of the cardboard. Remove the cardboard and hang it from another of its corners, using the weighted string. Make another mark along the string. Repeat this step one more time, making three lines across the cardboard. Where the lines cross is the center of gravity of the cardboard.

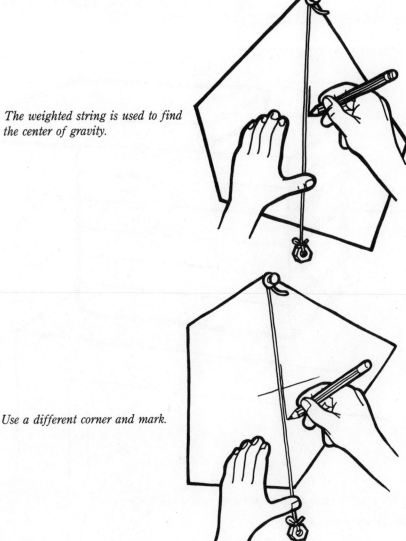

The weighted string is used to find the center of gravity.

Use a different corner and mark.

Experiment 33

Moving the Center of Gravity

Materials
- ☐ about 7 books of the same size
- ☐ table

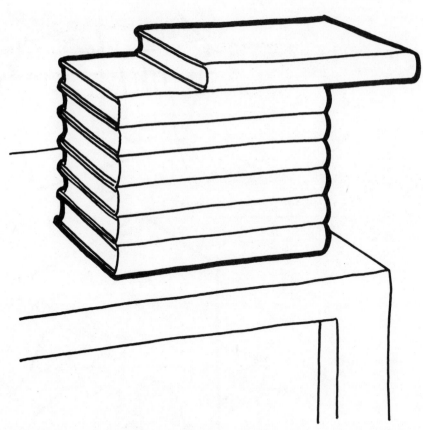

The center of gravity of the top book is over the table.

Stack the books neatly at the edge of the table. Slide the top book over the edge until it begins to tilt. Notice this is about halfway. Push it back a little so it rest solidly. Working from the top down, adjust each book over the edge of the table.

When you reach the bottom, the top book will be completely over the edge of the table. The top book can be moved the farthest. The next book is moved less because its center of gravity is affected by the book above it. The third book is moved even less as it has the two books above it affecting its center of gravity. The movement of each book is less as you work toward the bottom because the center of gravity of the books is affected by each book above it.

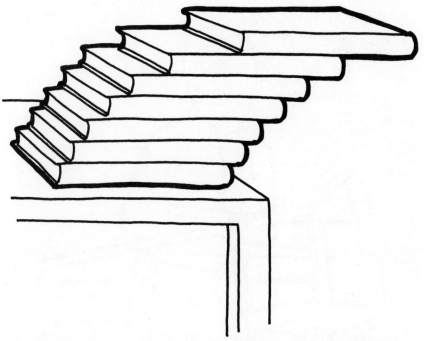

Now the center of gravity of the top book is well passed the edge of the table.

Experiment 34

Pendulum

Materials

- ☐ string (about 2 ½ feet long)
- ☐ weight (lead fishing sinker)
- ☐ 2 chairs
- ☐ broom
- ☐ book

Use a broomstick as support for the pendulum.

Place the chairs about 3 feet apart with their backs facing each other. Put the broom across the tops of the chairs to form a support beam. Next, tie one end of the string tightly around the middle of the broom handle, with the knot pointing straight down. Tie the sinker to the other end of the string so that it is slightly above the floor. Snip off any excess string at the sinker. When the sinker is pulled to one side and released, the weight will swing back and forth at a regular rate. The time it takes for the weight to go out and back is called the pendulum's period.

Stand up a book for a reference point.

Stand the book upright to one side as a reference point. Bring the weight close enough to touch the book and release it. The weight will swing out and back but will never come close enough to touch the book again.

When the weight is pulled to one side, gravity causes it to have potential energy. When the weight is released, the energy is changed to kinetic as the weight swings through its arc. At the end of its swing, the weight pauses for a moment and the kinetic energy becomes potential again. This change of energy is a continuous change, as the pendulum falls from its maximum height to its minimum height (its minimum speed to its maximum speed). This would continue forever if it were not for the friction and resistance of the air. This resistance is why the sinker didn't come back far enough to touch the book.

Experiment 35

Pendulums

Transferring Energy

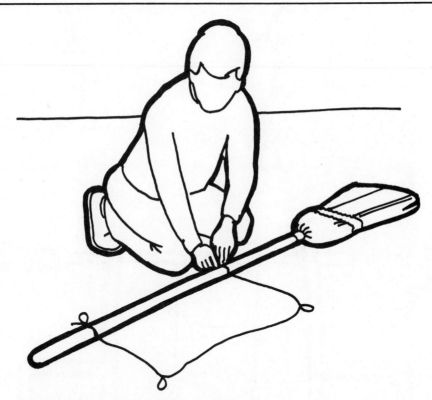

The string with two loops for two pendulums.

It is easier to assemble this experiment on the floor. First cut the string in half. Tie small loops about 12 inches from each end of one of the strings. Tie the ends of this string to the broom handle, about 2 feet apart. Cut the remaining string in half and tie one string

Suspend a weight from each loop.

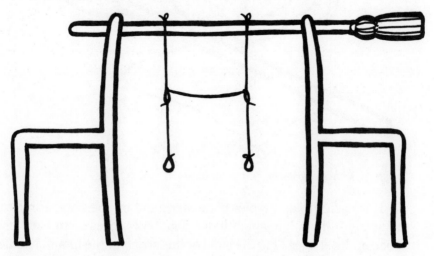

Place the broom stick on the chairs.

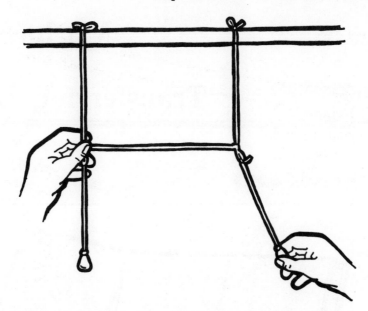

One pendulum is held by the loop, while the other is started to swing.

to each weight. Suspend each weight to the loops in the string tied to the broom. Place the chairs about 3 feet apart and carefully put the broom across the top of their backs.

After the weights have settled down, hold one still by holding it up by the loop and start the other weight swinging in a small arc, about 3 or 4 inches. Now release the loop of the other weight. At first both weights will swing back and forth. Then one will almost stop while the other one swings. But after a few swings it nearly stops and the first one starts to swing. This will continue for several minutes as the energy from the weights move back and forth through the string.

Experiment 36

Marbles

Transferring Energy

Materials

☐ about 8 marbles
☐ 2 or 3 books (hard back books about 1 inch thick)

A grooved ramp is built with books.

Lean the books against a wall with the backs down, making a track on top for the marbles. Prop one end of the books up about 1 inch to make a ramp. Keep the top level with the next book. Place the marbles on the level part of the track. Have all of the marbles touching each other. Now take one of the marbles up the ramp and release it. When it strikes the row of marbles, one end marble will roll away. If two marbles are rolled down the ramp, two end marbles will roll away. Three marbles down the ramp and three of the end marbles will roll.

The energy of the marble rolling down the ramp moves through the row of marbles and moves the end marble. The energy of two marbles rolling down the ramp is enough to move two marbles on the end.

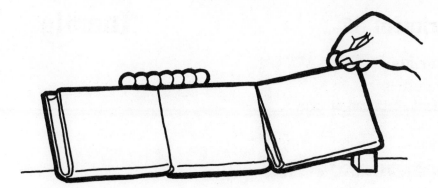

One rolling marble will cause the end marble to roll.

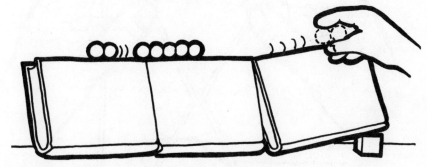

Two rolling marbles will cause the two end marbles to roll.

Experiment 37

Inertia

Materials

- [ ] short stick
- [ ] length of thread or light string
- [ ] brick

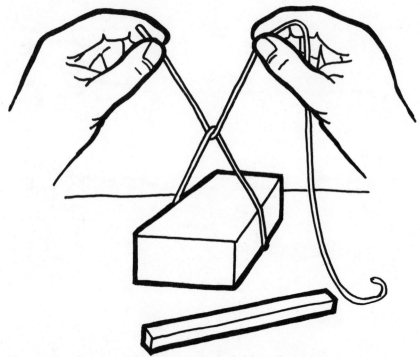

Use a strong thread or light string to tie the brick to the stick.

Tie one end of the thread around the middle of the brick and tie the other end around the middle of the stick. Using the stick for a handle, slowly raise the brick a few inches from the ground. Lower the brick back to the ground and jerk the string upwards. The string will break. The law of inertia states that any body at rest tries to stay at rest, while a body in motion will move in a straight line until it is affected by some outside force. It also takes a larger force to make a body at rest move abruptly than if it is done more gradual. In this case, the brick can be lifted if it is done gradually, but the brick resists the sudden change, and the thread cannot withstand the greater force so it breaks.

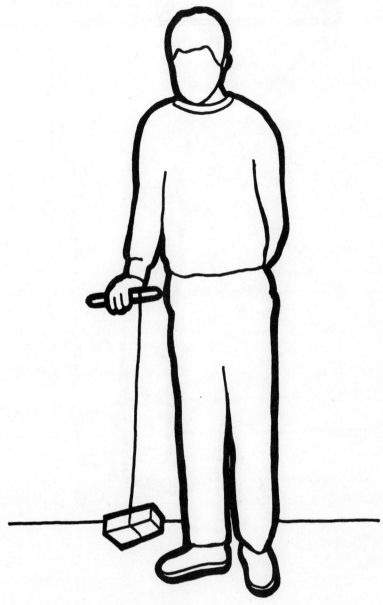

The brick can be lifted if its done slowly.

If the string is jerked it will break.

Experiment 38

Spinning Can

Materials

- ☐ hammer
- ☐ nail
- ☐ empty pop can
- ☐ running water
- ☐ length of string

Using the hammer and nail, carefully punch about four holes of equal distance around the bottom of the can. As you remove the nail from each hole, push the nail to one side. Push the nail in the same direction each time, making all of the holes point in the same direction. Bend the tab at the top of the can straight up and tie one end of the string in the tab opening. With the other end of the string, lower it under the kitchen faucet and fill the can with water. Lift the can with the string and it will quickly begin to spin as the water pours through the holes.

For every action there is an equal and opposite reaction. In this case, the water shoots out of the can at an angle and because the can is suspended by a string offering very little resistance, the force of the flowing water causes the can to rotate.

Make the holes on the side near the bottom.

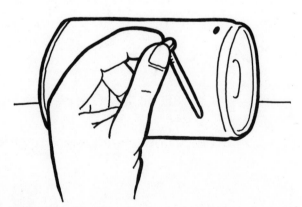

Bending the hole to the side will make the water shoot out at an angle.

The lift tab on the top of the can makes an excellent lift point.

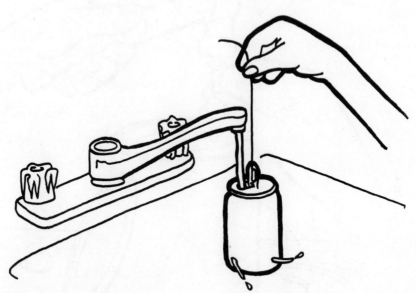

When the can is lifted by the string, the water flowing from the holes will start it to spin.

Experiment 39

Returning Can

Material

- ☐ coffee can with lid
- ☐ 4 match sticks or toothpicks
- ☐ 2 rubber bands
- ☐ weight (fishing sinker or washer)
- ☐ hammer
- ☐ nail

Punch two holes in each end of the coffee can.

Using the hammer and nail, carefully make a hole about one-third of the way across each end of the can and another about two-thirds. Thread the end of one rubber band through one of the holes in the bottom of the can. Fasten this end with one of the toothpicks. Thread one end of the other rubber band through the other hole in the bottom and fasten it with a toothpick. Thread the other ends of the two rubber bands through the holes in the lid and fasten them with the two remaining toothpicks. Tie the two rubber bands together around the middle of the can and attach the weight to this point. The rubber bands will be forming an X inside the can with the weight tied to the center. Replace the lid.

Roll the can on a smooth surface. It will travel a few feet, pause and then return. You might want to try different sizes of rubber bands and weights to obtain better results.

Rubber bands have a property called elasticity. This means that a rubber band can stretch and twist under a force and when the force is removed, it will almost return to its original shape. Inside the can, the weight and gravity create the force that twist the rubber band. The energy used to push the can will be stored in the twisted rubber band. When the can stops, this stored energy will cause the rubber bands to unwind and return the can near its original position.

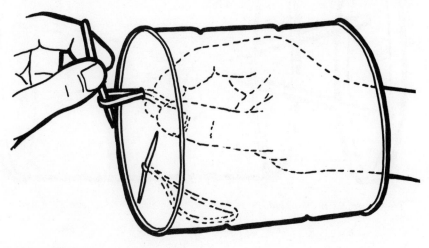

Hold the rubber bands by toothpicks.

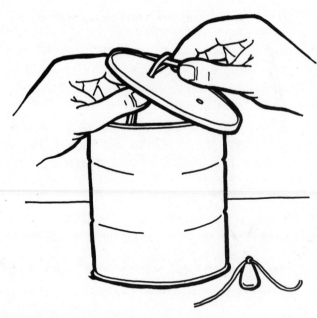

Attach rubberbands to lids.

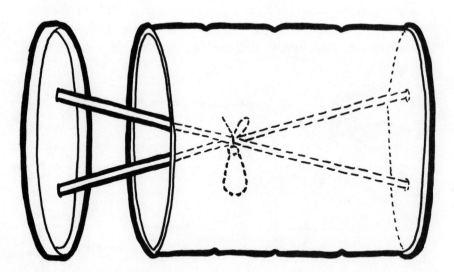

Tie the weight in the middle.

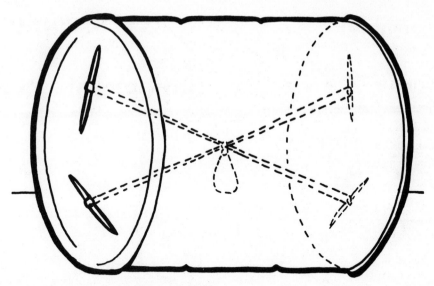

Inside view showing completed assembly.

Experiment 40

Gyroscopic Phonograph Record

Materials
- ☐ length of string (about 3 or 4 feet)
- ☐ toothpick or small stick
- ☐ phonograph record

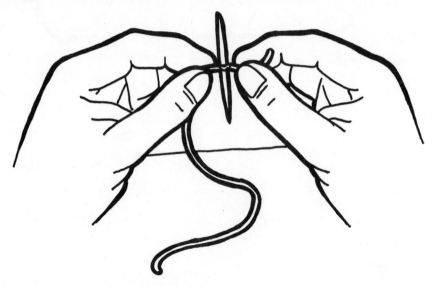

Use a toothpick to support the phonograph record.

Tie one end of the string around the middle of the toothpick and thread it through the hole in the record. Hold the other end of the string and suspend the record a few inches above the floor. Start the record to swing slowly from side to side like a pendulum. Notice how it tilts around as it swings back and forth. Now try it again, but this time hold the record level and give it a spin. This time the record seems to float above the floor keeping its position level. It will continue to stay level until the record almost stops spinning.

Thread the string through the hole in the record.

The record will hang straight up and down.

The record has become a gyroscope. As long as it is spinning it will remain in the same plane, or attitude it has, when it began to spin. This is why a spinning top will stand on its point and the rotating wheels of a bicycle keep it upright.

Start the record spinning.

The record will swing level with the floor.

Experiment 41

Inertia and Centripetal Force

Materials
- ☐ 4 feet of strong string
- ☐ tennis ball

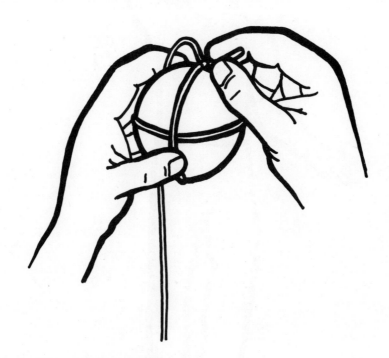

Tie the string securely to the ball.

Tie one end of the string securely around the tennis ball. Holding the other end of the string, start the ball swinging in a circle around your head. Not too fast, just enough to keep the ball and string level. As long as you hold the string and continue to swing, the ball will travel in an orbit around your head.

As the ball travels around, its momentum makes it try to pull away from the center of its orbit. If the string breaks, the ball will

Centrifugal force tries to pull the ball away while centripetal force pulls the ball in.

fly away. This is caused by inertia. But the string is creating another force that tries to pull the ball in toward the center of the orbit. This is called centripetal force. It is balanced against the inertia and keeps the ball in orbit.

Experiment 42

Conservation of Energy

Thread one end of the string through the opening in the weight and tie securely with a knot. Feed the other end of the string through the hole in the spool. Find a clear place outdoors with no one standing nearby.

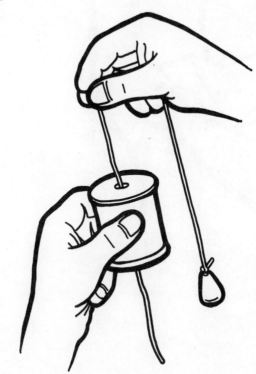

A spool is used so that the string will slide easily through your hand.

Hold the end of the string that does not have the weight tightly in one hand, and hold the spool over your head with the other hand. Start the weight swinging in a large circle.

Swing the weight in a large circle.

Try to keep the weight circling at a nice steady rate. Notice the speed the weight is traveling. Hold the spool at the same height and pull down on the end of the string. As the weight moves in toward the spool it will speed up. This happens because, when the weight was traveling in a larger circle, it was moving at a steady speed representing a certain amount of energy. When the circle became smaller, the weight tried to maintain the same speed and energy. The smaller circle meant the distance the weight had to travel would be less. In order to keep the same energy level, it had to make more revolutions, or orbits, for the same amount of time.

As the size of the circle gets smaller, the weight will make faster orbits.

Experiment 43

The Law of Falling Bodies

Materials

- ☐ softball (small, rubber ball)
- ☐ golf ball
- ☐ 2 sheets of paper the same size
- ☐ high platform (upstairs window or porch)

A softball and a golf ball dropped at the same time will hit the ground at the same time.

Make sure the area is clear below you. Hold the balls side by side and release them at the same time. Both balls will hit the ground at the same time even though the golf ball is lighter. The force of gravity pulls on all bodies the same, regardless of shape, size, or weight.

Air resistance will affect the sheet of paper more than the wad of paper.

Crumple one of the sheets of paper into a ball and drop both sheets of paper at the same time. The ball of paper will fall much faster, even though it is the same weight as the flat sheet of paper. This happens because of the resistance of the air against the falling bodies. A feather and a cannon ball will fall at the same speed in a vacuum.

If a ball is thrown perfectly level with the ground the same time another ball is dropped from the same height, the thrown ball will travel farther horizontally, but they both will hit the ground at the same time. The horizontal motion does not change the rate of fall of the thrown ball. It is traveling, but it is also falling. And it will fall at the same speed as the ball that was dropped.

Experiment 44

Parachute

Attach strings to each corner.

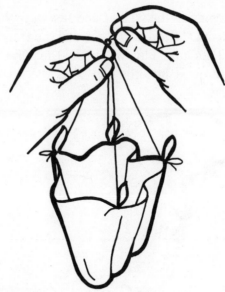

Tie the weight to the strings.

The folded parachute offers less resistance to the air.

Tie one end of the strings to each of the corners of the handkerchief. Suspend the handkerchief from its center and bring the four strings together so their lengths are equal. Thread the ends through the opening in the weight and fasten with a knot. Fold the handkerchief, from the top down, toward the weight, wrapping the strings around the cloth. This will make the parachute into a small bundle. Now, throw it into the air.

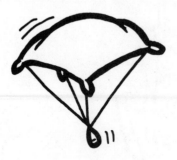

The opened parachute has much greater resistance to the air.

The parachute bundle will travel up until it unfolds. Then the parachute will open allowing the weight to come down slowly. When the parachute goes up it is a small bundle and the air resistance is little. When the parachute opens, however, the air resistance is greatly increased, slowing the descent.

Experiment 45

Helicopter

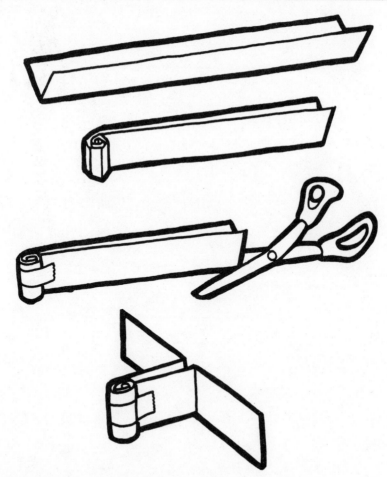

A simple helicopter can be made from a strip of paper.

Fold the paper in half lengthwise. Make about 10 small bends on one end for weight. Fasten these bends with a small piece of scotch tape. On the other end, cut down the center of the fold about 4 inches and bend the two strips outward to form narrow wings.

Drop the helicopter from a high platform or from above your head and it will rotate, slowing its descent to the ground. The weighted end creates the center of gravity and air flowing past the narrow wings cause them to rotate, slowing its descent. Helicopters are called rotary-wing aircraft.

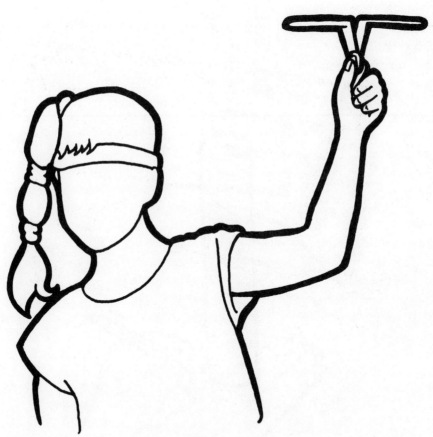

When the helicopter is dropped, it will spin rapidly as it descends.

Experiment 46

Propeller Test Stand

Materials

- ☐ paper clip
- ☐ 1 empty cardboard tube from a roll of paper towels
- ☐ scotch tape or glue
- ☐ empty cereal box
- ☐ scissors
- ☐ toothpick
- ☐ button
- ☐ rubber band
- ☐ string or wire hook

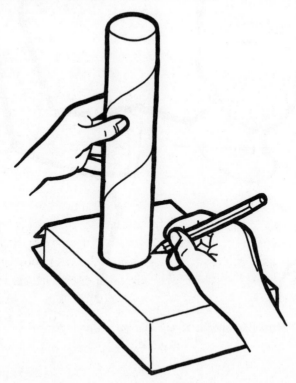

Mark the circle for the cover for the end of the tube.

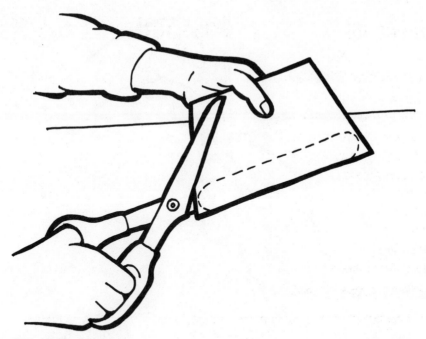

Cut out the propeller.

Using the end of the tube as a guide, mark and carefully cut out a circle from the cereal box. Make a small hole in the center of this circle. Cut a propeller from the cereal box about 5 inches long and ½ inch wide. Make a small hole in the center of the propeller. Straighten the paper clip and stick one end through the hole in the propeller and bend the end about ½ inch to form a sharp corner. Tape or glue this flat bend to the propeller. This fixes the propeller to the paper clip propeller shaft. Thread the other end of the paper clip through one of the holes in the button and through the hole in the cardboard circle. Bend a small loop in the end of the paper clip and attach one end of the rubber band. Using a string or wire hook, stretch the rubber band through the tube and loop around the toothpick. Place the toothpick across the end of the tube. The toothpick and the cardboard circle at the other end can be held in place by tape or glue. Make a twist in the propeller blades.

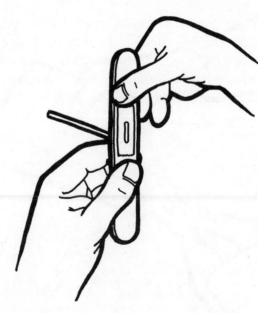

Glue or tape the propeller to the shaft.

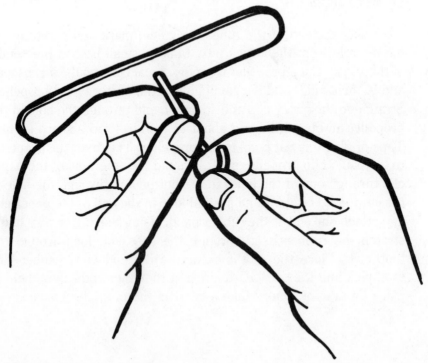

Bend a hook for the rubber band.

Holding the tube in one hand, wind the propeller with one finger of the other hand. You might have to wind for a minute. When you release the propeller, the twisted rubber band will unwind, spinning the propeller and pushing the air in the direction determined by the twist in the propeller. This force of air is called thrust and provides the energy to drive boats and airplanes.

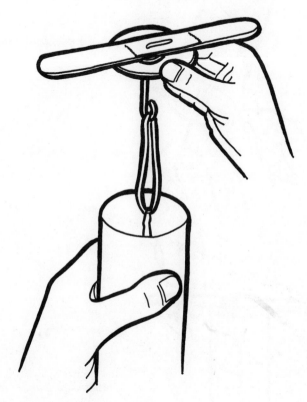

Lower the rubber band through the tube.

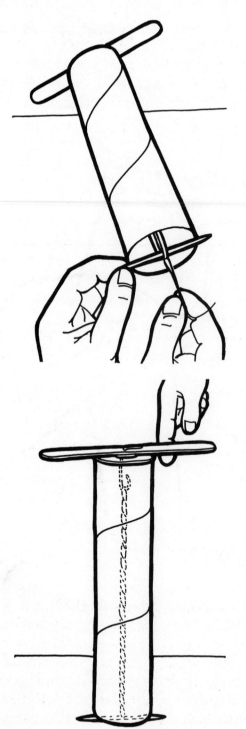

Secure the rubber band with the toothpick.

Inside view showing completed assembly.

Experiment 47

Airfoil

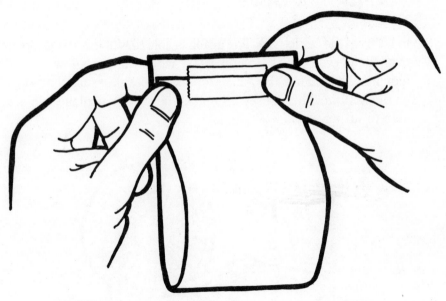

Use scotch tape to fasten the ends together.

Fold the paper in half so it is about 5 inches long. Hold the ends together and tape it to the card. Before sticking the tape, however, move the end of the top of the paper so that the paper opens slightly into the loop. This makes the bottom of the loop flat with the cardboard, and the top half curved. Stick the airfoil to the cardboard Hold the cardboard in front of the fan with the tape ends pointing toward the fan. Turn the fan on high and the paper airfoil will try

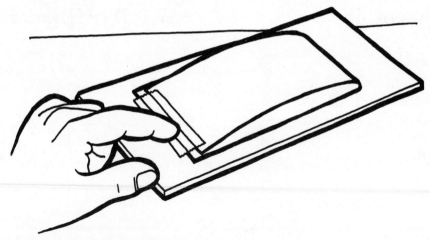

Attach the airfoil to the cardboard.

to lift from the cardboard. The airfoil is able to lift because air flowing over the curved top is traveling fast while the air below is hardly moving. This makes the air on top exert less pressure than the air on the bottom. This causes lift and is the reason airplane wings are able to lift an airplane.

The current of air will try to lift the airfoil.

Experiment 48

Paper Airplane

Materials
☐ drinking straw
☐ sheet of paper
☐ scotch tape
☐ scissors

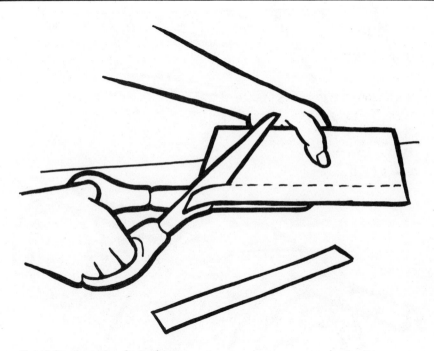

Cut strips from the sheet of paper.

Carefully cut two strips about ½ inch wide and 11 inches long from the sheet of paper. Make a ring with each strip and tape the ends together. Tape one ring to the side of the straw about two inches from one end. This is the front. Tape the other ring to the same side of the straw about one inch from the other end. Toss the airplane into the air. With the rings located in this position, it will have a slow flight path. Repositioning the rings will change the flight path.

Form the strips into paper rings.

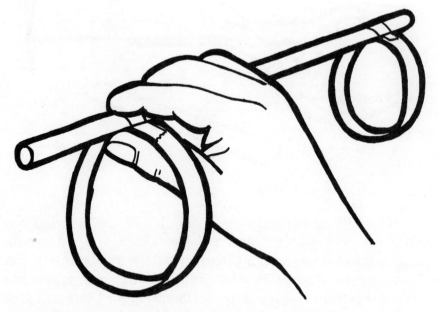

Use scotch tape to fasten the rings to the straw.

Hand launch the airplane in an open area.

Experiment 49

Materials
- ☐ large spool of thread
- ☐ table

With the end of the thread elevated, the spool will roll away.

Unwind about 2 feet of thread and place the spool on its side on the table. Have the thread coming from the side next to the table. Hold the end of the thread about a foot above the table and pull on the string. The spool will roll away, unwinding more thread. Lower the end of the thread to about an inch above the table and try again. This time the spool will roll toward your hand and try to wind up the thread. When the end of the thread is pulled from the higher angle, the force is being applied to the spool from the same side. But when the thread is pulled from the lower position the force is applied to the opposite side of the spool.

With the end of the thread down, the spool will roll toward you.

Part III

SCIENCE FAIR PROJECTS

Probably the biggest problem with a science fair project is choosing a subject. Don't be in any hurry, spend a little time picking the subject. This is the planning stage. It might be the most important part of the project.

Deciding on a subject requires a lot of thought.

You will have to use some imagination to build on the experiments that are given in this book, but the basic principles are there.

Divide your science fair project into steps. For example:

(1) choosing a topic

(2) questions and hypothesis. A hypothesis is simply what you think will happen in the experiment

(3) doing the experiment

(4) results and conclusions.

A report on the experiment is usually necessary. It should show the purpose of the experiment. The purpose should answer a question or prove a hypothesis. The report should include the experiment

Keep reports on your project.

itself, the results of the experiment and the conclusions that were made. Graphs and charts can be used.

Choose a topic that you really want to learn about. A science fair project should be fun as well as educational. Don't make your project too complicated. Choose a topic that you can complete with the materials and equipment you have available or can build. Some of the greatest scientific discoveries were based on a simple principle.

After selecting a topic, break it down to a specific problem to solve, or a question to be answered. For example, the subject of friction could be narrowed down to comparing the friction of different surfaces such as sandpaper, cardboard and waxed paper. This can be done by taping strips of the different materials to a flat surface and measuring the force required to pull a block of wood across each surface.

An experiment on static versus moving friction could be demonstrated by a belt, similar to a conveyer belt, turned by a hand crank. A weight, such as a short piece of 2 × 4 inch lumber, could be connected by a string to a small spring scale. When the crank

A project can often be built using items found around the home.

is turned, the reading on the scale can be noted. As the crank is continually turned, the comparative reading on the scale noted.

Suppose you are interested in the experiment with the needle floating on water; you could show that a single drop of detergent can reduce the surface tension on a number of square inches, enough to sink the needle. Surface tension will also support a double-edged razor blade. You might compare the two.

The homemade barometer could be another topic for a science fair project. This can be built and calibrated with your local weather report and then monitored for several days to compare its accuracy.

Once you have selected a topic, think about how you're going to display your project. A model will probably be necessary. Most can be assembled from wood and cardboard. Don't overlook normal household throw aways; empty jars, cardboard tubes from paper towels, and empty coffee cans. Try to be original. Use your imagination and be creative.

Suppose you wanted to show why air is used in some shock absorbers on automobiles. The diving eye dropper experiment will prove that air is easier to compress than water. You could use posters to display drawings showing water or brake fluid inside a cylinder, compared to another cylinder with air in it. The air would compress, making a smooth ride, while the other one would not.

Squeezing the plastic bottle of water that contains the eye dropper can demonstrate this fact. Air makes a good shock absorber. In addition, by increasing or decreasing the air pressure inside the shock absorber, you can adjust the softness of the ride.

The project can be displayed on a table in front of a self-supporting panel. The panel can be made from heavy cardboard cut in three pieces. One larger piece would be the back, while two smaller pieces, angled slightly forward, would be the sides. It would look something like a miniature theater stage.

On the left side you could show the purpose of the experiment. The middle of the panel would contain drawings of the experiment, and the right side of the panel can show the results and conclusions of the experiment.

A cardboard panel can be used to display the progress of your experiment.

Try to be creative and develop your own ideas rather than simply copying them from a textbook. The experiment doesn't have to be original, just try it from a different point of view. Ask yourself what if I did it this way instead. Powered flight has been around for some time but because of new ideas from different experimenters, aviation has made dramatic advances.

Index

Other Bestsellers of Related Interest

EXPLORING EARTH FROM SPACE—Jon Erickson

Learn how orbiting satellites are used to explore our planet. Geophysicist Jon Erickson covers the technology—how satellite images are collected and processed—and describes how this technology is applied in weather forecasting, land-use planning, geologic mapping, mineral exploration, agriculture, disaster control and more. 207 pages, 157 illustrations. Book No. 3242, $15.95 paperback, $23.95 hardcover

UNDERSTANDING LASERS—Stan Gibilisco

If you could have only one book that would tell you everything you need to know about lasers and their applications—this would be the book for you! Covering all the different types of laser applications—from fiberoptics to supermarket checkout registers—Stan Gibilisco offers a comprehensive overview of this fascinating phenomenon of light. He describes what lasers are and how they work, and examines in detail the different kinds of lasers in use today. 180 pages, 96 illustrations. Book No. 3175, $14.95 paperback, $23.95 hardcover

SUPERCONDUCTIVITY—The Threshold of a New Technology—Jonathan L. Mayo

Superconductivity is generating an excitement not seen in the scientific world for decades! Experts are predicting advances in state-of-the-art technology that will make most exisiting electrical and electronic technologies obsolete! This book is one of the most complete and thorough introductions to a multifaceted phenomenon that covers the full spectrum of superconductivity and superconductive technology. 160 pages, 58 illustrations. Book No. 3022, $12.95 paperback, $18.95 hardcover

THE TOOLS OF SCIENCE: Ideas and Activities for Guiding Young Scientists—Jean Stangl

Providing science experiences that will stimulate children's curiosity and motivate them to explore and discover the world around them is a challenge for adults today. This book is an excellent course for teachers, parents, and youth leaders looking for fresh ideas. Jean Stangl has taught both children and teachers for over 25 years. This extensive experience is evident in this delightful book that will create hours of educational fun for elementary school children. 160 pages, 26 illustrations. Book No. 3216, $8.95 paperback, $16.95 hardcover

THE LIVING EARTH: The Coevolution of the Planet and Life—Jon Erickson

The latest in TAB's Discovering Earth Science Series, this book explores the origin and evolution of life, including the geological and biological areas. The author covers the ice ages, the geological eras, the beginnings of life, dinosaurs, plants, and more. Discussions of extinction and the destruction of life throgh erosion, pollution, and other hazards are also presented. This is yor opportunity to discover the variety and complexity of the Earth's biosphere. 208 pages, Fully illustrated. Book No. 3142, $14.95 paperback, $22.95 hardcover

THE MYSTERIOUS OCEANS—Jon Erickson

Explores far below the foamy crest and delve into the wonders of the sea—its forces, its predators, its role in the food chain, its mountains, and much more. The author covers topics of oceanography, geology, meteorology, and marine biology. 208 pages, 169 illustrations. Book No. 3042, $14.95 paperback, $22.95 hardcover

---

Look for These and Other TAB Books at Your Local BOOKSTORE
